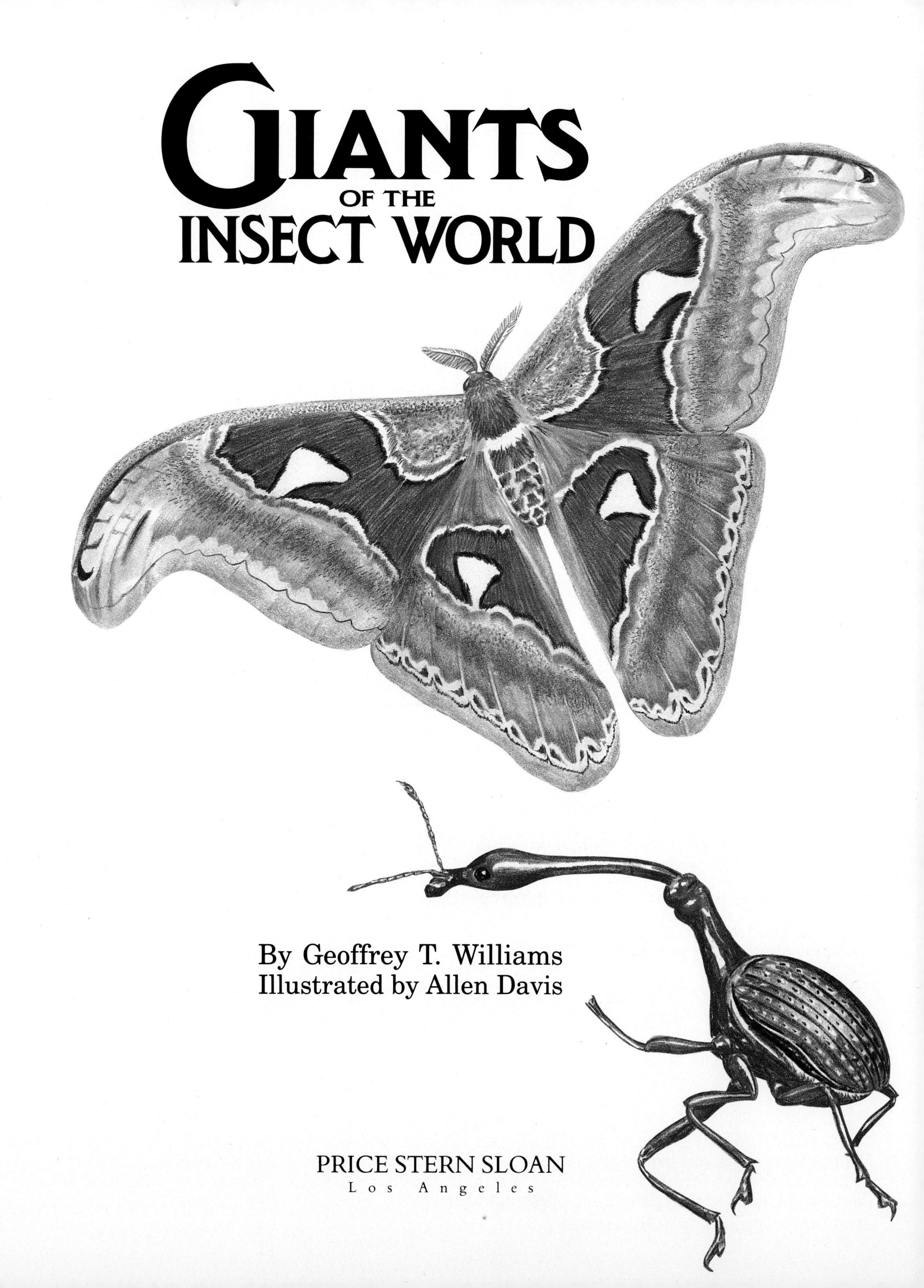

GIANTS OF THE INSECT WORLD

By Geoffrey T. Williams
Illustrated by Allen Davis

PRICE STERN SLOAN
Los Angeles

Acknowledgments

The author wishes to thank
David K. Faulkner, Entomologist, Natural History Museum, San Diego, California;
Dr. Robert Gordon, Entomologist, National Museum of Natural History, Smithsonian Institution;
and for her valuable time and assistance in technical editing,
I especially wish to thank Sally H. Love, Director, The Insect Zoo,
National Museum of Natural History, Smithsonian Institution.

About the Author

Geoffrey T. Williams is the author of *Adventures in the Solar System, Beyond the Solar System, Hello, Mars!, Treasures of the Barrier Reef, Dinosaur World, Explorers in Dinosaur World, Lost in Dinosaur World, Saber Tooth* and *The Alien Next Door.* He is also the producer and sound designer of the audio cassettes that accompany his Dinosaur books. Geoffrey lives in northern San Diego County with his wife, daughter, two birds and a very old Manx cat.

About the Illustrator

Allen Davis has been drawing since he could hold a pencil. Educated at the Parsons School of Design, he won the "Outstanding Science Books for Children" award in 1979 for his first picture book. His work appears in many young-adult novels, anthologies and the Picture Book Art Gallery in Chatham, Massachusetts. He lives in New York with his wife and son.

Published by Price Stern Sloan, Inc.
Los Angeles, California

Printed in Hong Kong

10 9 8 7 6 5 4 3 2 1

First Printing.

Library of Congress Cataloging-in-Publication Data

Williams, Geoffrey T.
Giants of the insect world/by Geoffrey T. Williams.
p. cm.
Summary: A boy travels around the world with his parents, a trip which enables him to see and collect unusual insect specimens.
ISBN 0-8431-2832-1

1. Insects—Juvenile literature. [1. Insects.] I. Title.
QL467.2.W55 1991
595.7—dc20 90-25800
CIP
AC

This book is printed on acid-free paper.

Contents

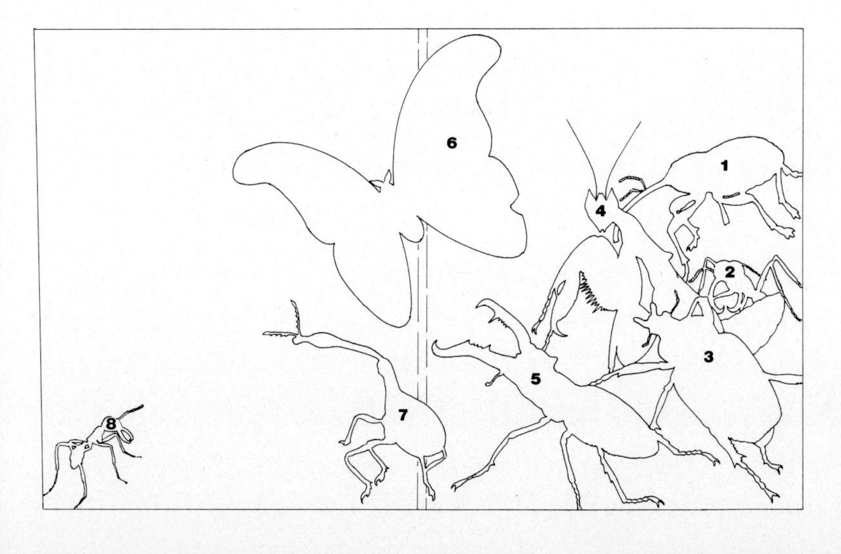

KEY TO THE INSECTS ON THE COVERS

1. ELEPHANT WEEVIL
2. ARMY ANT
3. GOLIATH BEETLE
4. MALAYSIAN FLOWER MANTIS
5. EUROPEAN STAG BEETLE
6. ATLAS MOTH
7. GIRAFFE WEEVIL
8. ARMY ANT

Introduction

Some kids like baseball—they can take batting practice and shag fly balls all day long; some kids like video games—they can play *Blastron* or *Strike Force* until their eyes fall off.

With me it's insects. Ever since Dad gave me my first ant farm, and ever since I collected my first butterfly, I've thought insects were great.

Don't get me wrong, I like ball games too. Mom and Dad and I go all the time. And I'm the *Blastron* champion of the fifth grade. But give me a good bug hunt any day.

Entomologists, the scientists who study insects, spiders and animals like that, say there are more different kinds of insects than there are all the other plants and animals in the whole world put together—almost a million! So I'll never run out of new ones to add to my collection. Of course, since a lot of the neatest insects live in places thousands and thousand of miles away I didn't think I'd ever have a collection as good as the ones I saw in museums.

But one day, just before school let out for the summer, Dad came into my room to talk about vacation.

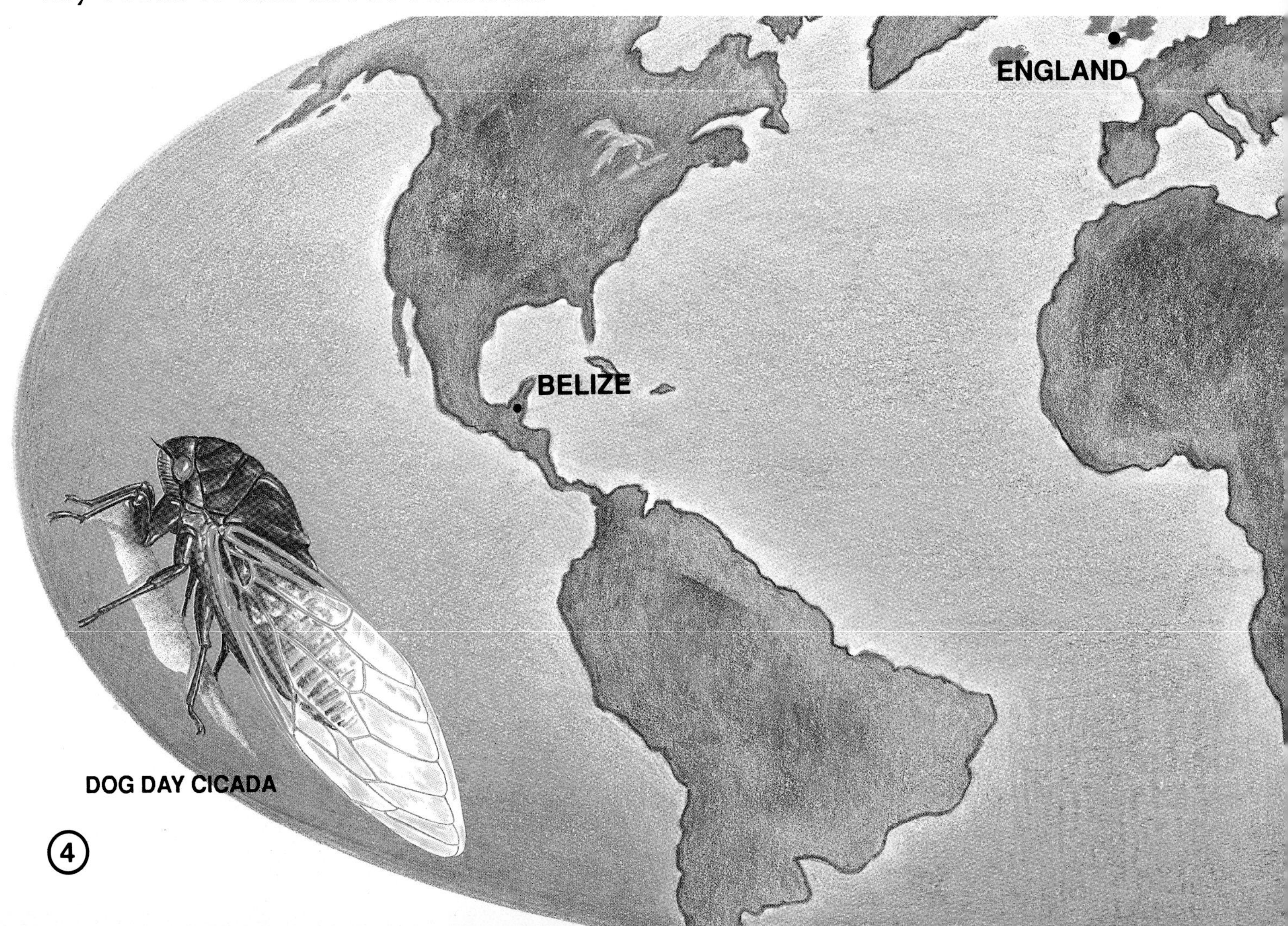

I was mounting my new **dog day cicada** *(Order Homoptera)*. I'd been out collecting when I'd heard this singing sound—like a cricket makes but a lot louder—and the next thing I know this buzz bomb is coming right at me: four-inch wings, thick body, big, bulging brown eyes. I ducked, stuck up my net, and got lucky: in he went. (I knew it was a "he" because only the male cicadas sing to attract mates.) Then I put him in my collecting jar, which painlessly kills insects so I can mount them for my collection.

"Max, we're getting ready to take a little trip...."

"Great! Are we going to the mountains again?" Mountains are a good place to hunt insects.

He gave me a funny smile. "Not exactly. We're going around the world."

My mouth fell open and I stared at him. "Around the world? When?"

"As soon as school's out. We'll be going to Australia, Malaysia, Japan, Africa, Europe and Central America."

For an insect collector like me it sounded like the Superbowl, World Series and N.B.A. Playoffs all rolled into one. "Dad, those are places where some of the biggest insects in the world live!"

He grinned, "Then I guess we'll just have to make room for your collecting equipment and camera."

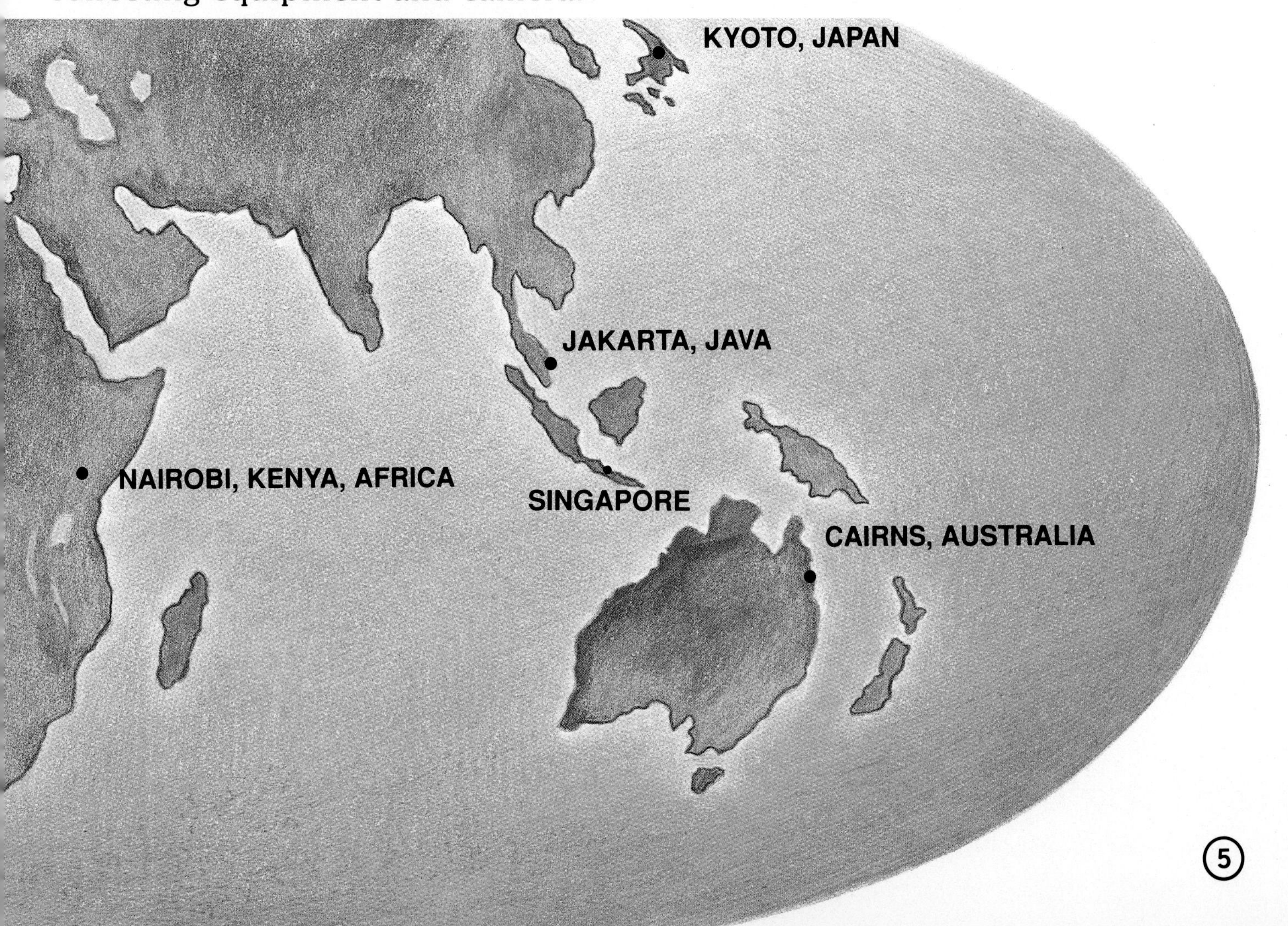

What's an Insect?

Why am I so interested in insects? Lots of reasons, I guess. Some insects are really helpful. Like honeybees. Without honeybees to pollinate plants a lot of crops wouldn't grow. Some insects are dangerous. Mosquitoes can carry a deadly disease called malaria. And maybe it sounds funny, but I think a lot of insects are beautiful—moths, butterflies, crickets, most beetles and some of the weird-looking ones like walkingsticks and mantises.

Insects belong to the phylum *Arthropoda*. A phylum is a large group of animals with similar characteristics. Arthropod means "jointed foot." Humans belong to the phylum *Chordata*—animals with spinal chords.

You might not know it, but I'll bet you eat animals that are a lot like insects. Crabs, shrimp, crawdads and lobsters are all Crustaceans—a kind of Arthropod—and first cousins to insects. Lots of people think spiders are insects too. They're not. They're another kind of Arthropod called Arachnids. I don't know anybody who eats Arachnids.

How do you tell if an insect is an insect and not some other kind of creature? Well, if you see something crawling up your leg that has three body parts—head, thorax and abdomen—one pair of antennae, three pairs of jointed legs located on the thorax (the middle body part), and two pairs of wings, it's an insect. Crustaceans usually have more legs and most of them live in the water. Spiders always have four pairs of legs but only two body parts.

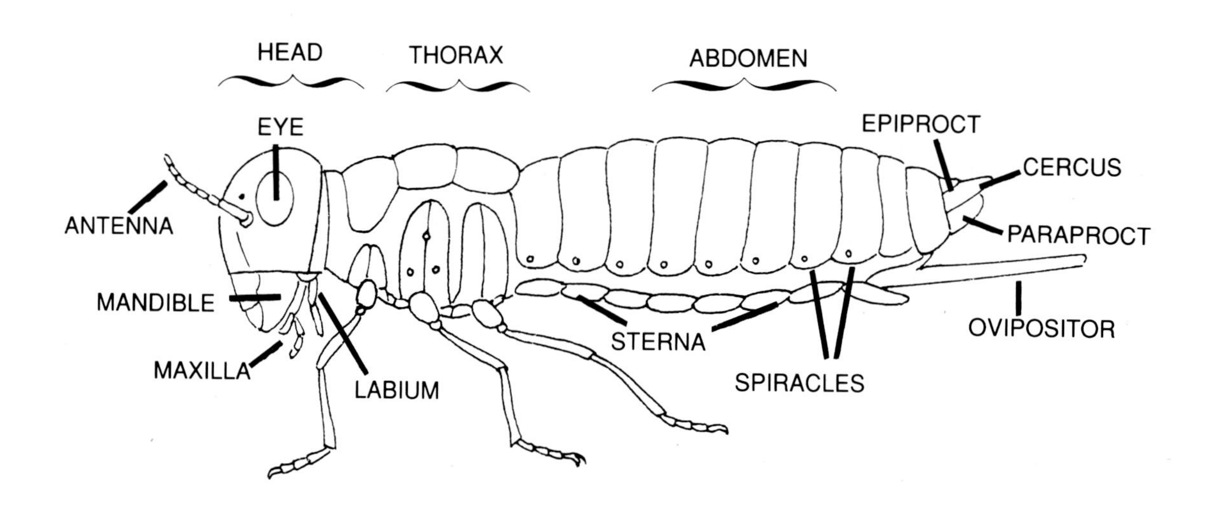

DIAGRAM OF AN INSECT

Insects are great survivors. They were the first animals that learned how to breath out of water and walk on land. They were around, starting about 200 million years before dinosaurs, and they're still here, millions of years after the last Tyranasaurus Rex became extinct.

Insects have all sorts of special things to help them survive. First of all, most insects have wings to escape from predators (other animals that might eat them). And insects don't have bones that can break. Instead, they have a hard outer skeleton called an *exoskeleton* that helps protect them.

Insects' muscles are attached to their exoskeletons in a way that makes them very strong for their size. I remember a picnic a few years ago during which my folks and I watched ants carry off huge crumbs of food.

"Ants can carry more than fifty times their own weight, Max," Mom told me. "If you were that strong you could carry our car home."

I thought that was pretty cool until Dad said, "There are some big beetles that can lift more than eight hundred and fifty times their own weight. One of them might be able to carry you to its home, Max." I gulped. "But don't worry," he said, laughing, "you're much too big to fit through its front door."

Insects have lots of ways of protecting themselves. Some use poison. Blister beetles have a chemical that can blister your skin if you touch them. Some use camouflage, a way of disguising themselves so they're hard to see. The Kallima butterfly can look exactly like a dead leaf on a tree branch. Some insects even protect themselves with smell. The shieldbug can make a terrible stink if handled or attacked—that's why some people call them stinkbugs.

Insect legs are used for everything from feeding and tasting to digging and fighting and swimming.

SIDE VIEW OF MOSQUITO HEAD AND MOUTH

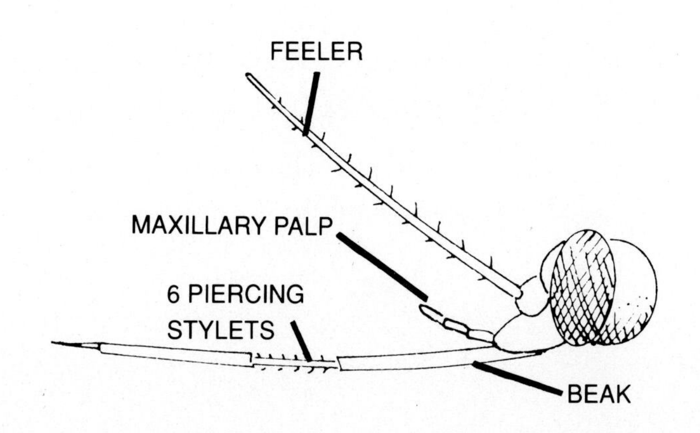

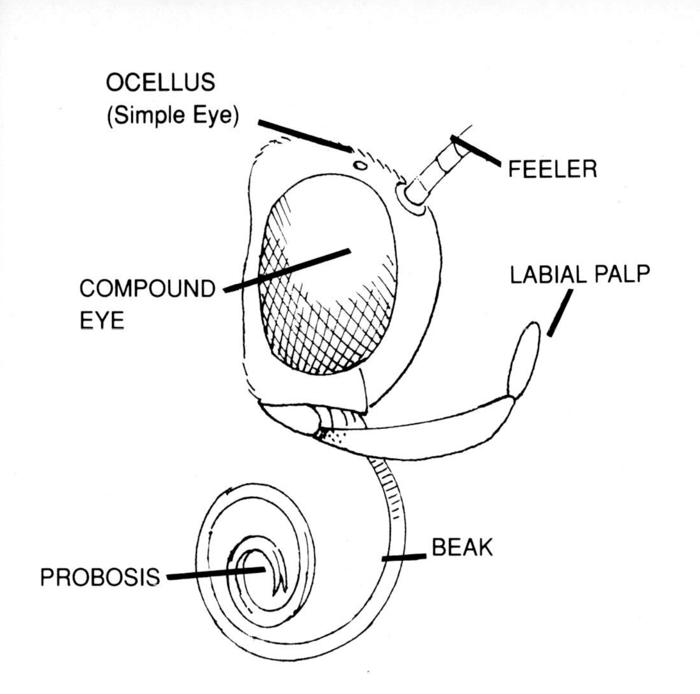

SIDE VIEW OF MOTH HEAD AND MOUTH

MONARCH CATERPILLAR

Insects have different kinds of mouths: mosquitoes' mouths are made to pierce skin and suck blood; butterflies and moths have mouths made to drink nectar from flowers; praying mantises' mouths are made to chew other insects.

Most insects have two different kinds of eyes. Two or three small, simple eyes called *ocelli*, which can't see much except the difference between light and dark, and two big *compound eyes* that they use to see everything else. They're called compound eyes because they're made of thousands of separate lenses. Humans have only one lens on each eye.

Another big reason insects have survived millions of years is that they reproduce in huge numbers: A queen honeybee can lay as many as two thousand eggs a day and the queen of an East African termite species lays over forty thousand eggs a day! So, no matter how many get killed or eaten, there's always more left over.

MONARCH BUTTERFLY CHRYSALIS

Insects grow by shedding their outgrown exoskeleton and growing a new one. This is called *molting.* Insects molt several times before they become adults. Each molt is called an *instar*. As insects grow, they change. Entomologists call this *metamorphosis*. Different insects metamorphose differently. Some insects hatch from eggs and are just small-sized versions of the adults. This is called simple metamorphosis. Other insects, including cockroaches, grasshoppers, mantids and termites, go through incomplete metamorphosis. They hatch from eggs and look like small-sized adults also, but each time they molt their wing pads grow larger.

Most insects undergo complete metamorphosis. They hatch from eggs too, but the young, called *larva*, usually look like worms and don't resemble the adults at all. Caterpillars, which are butterfly larva, usually go through four or five instars over several months. During the last instar, the caterpillar fastens itself to a tree branch by spinning silk threads. It molts one last time, shedding its old exoskeleton like crumpled paper. Underneath is a soft body called a *pupa* that hardens to become a *chrysalis.*

Then a really strange thing happens. The larva's fat and muscle tissues are broken down by chemicals into kind of an organic soup, the organs grow larger, the mouthparts change from chewing mandibles to a long probosis for sucking, wings develop and reproductive organs form. This process can take from several weeks to several months, depending on the type of butterfly. When the chrysalis splits open, out crawls a beautiful butterfly—an animal that doesn't look anything like the caterpillar it came from. Dad said this would be like him taking the car apart and turning it into a jet plane. It's all the same material—rubber, metal and plastic—but it's a whole new way of putting it together.

ADULT MONARCH

An Insect with Horns

Our first stop was Kyoto, Japan, a big, crowded city, full of neon lights and honking cars. But you would never know it where we were.

We were visiting Saihoji, a Japanese temple-garden more than a thousand years old. It seemed completely cut off from the busy city nearby. It was quiet and green, filled with moss and trees and still pools of water.

"It's so tidy," I said, looking around. I couldn't see a rock or a leaf out of place.

"The Japanese gardeners plan and work hard to keep it that way," Mom explained. "To them, these gardens are works of art."

"Well, there's something I'll bet they didn't plan on," Dad said, pointing to the ground in front of us.

Mom gasped, "My goodness! That thing has horns!"

Waddling across the stone path was one of the biggest insects in the world—a **rhinoceros beetle** *(Order Coleoptera)*.

"I'm glad we don't have any of those back home," she said.

"We will have soon," I told her as I picked it up and put it into my collecting jar. "Actually, Mom, there's a Hercules beetle back home with horns. But he's barely two inches long. This guy's almost twice as long."

"That must be the biggest beetle in the world. And it looks dangerous."

"It uses its horns only to fight another male for a female, Mom. But even then the horns aren't used to stab or stick, just to turn the other beetle over so it's helpless. And this isn't the biggest beetle in the world. The Goliath beetle in Africa is more than six inches long and weighs as much as a full-grown rat." Mom made a face and said, "Yuck," so I decided not to tell her that the natives thought the larva of the Goliath beetle tasted delicious when eaten raw or lightly roasted.

There are more different kinds of beetles than any other animal in the world—almost 300 thousand. The biggest is the Goliath. The smallest, the Hairy Winged Beetle, is so tiny it can walk through the eye of a needle. Beetles live everywhere on earth except Antarctica.

"Lots of beetles are destructive, aren't they, Max?" Dad asked.

"Yeah. I know the cotton-boll weevil has caused billions of dollars in damage to cotton. But some beetles are helpful."

"That's right," Mom spoke up. "The ladybug beetles in the garden eat the aphids that hurt my plants."

Birdwing

We took a huge jet plane south for thousands of miles to Cairns, a city on the northeast coast of Australia. From there we took a small plane to the mouth of the Endeavour River. Lots of people come here to scuba dive on the Great Barrier Reef, so for the first few days I went with Mom and Dad as they explored the beautiful underwater world.

But on the third day it was my turn.

"Where are you taking us, Max?" We were walking along the bank of the river. The air was heavy with the smell of living, growing things. There were strange forest sounds all around, and sunlight filtered through the leaves of the tall trees.

"Not too far, Mom. Maybe a mile or so."

"This thick undergrowth and all the mangrove trees remind me a little of the Florida Everglades," Dad said.

"What exactly are we looking for?" Mom asked.

"A butterfly."

"But you've got plenty of butterflies in your collection."

"Not like these guys."

We rounded a turn and came upon a quiet pool of water. I stopped still, almost afraid to breath. I couldn't believe it. There it was. "Look," I whispered.

"Oh, Max," Mom said, "it's gorgeous."

It was a **birdwing butterfly** *(Order Lepidoptera)*, the largest butterfly in the world. Its black velvet and bright green wings, almost a foot across, were moving slowly up and down as it drank from the water. Butterfly wings are actually thin, transparent skin criss-crossed with veins. The colors come from thousands of tiny scales so delicate they brush off like dust on your fingers. In fact the word lepidoptera means "scaled wings."

A butterfly's mouth is a long coiled tube called a *proboscis*. It is used for drinking and to suck nectar from flowers. Butterflies use their antennae for tasting and the organs on their feet for smelling.

I carefully got my camera out and took some pictures.

"Really something, isn't it?" a voice said behind us.

We turned around and saw a tall man in khaki pants and shirt, and a wide, flat-brimmed hat. He was carrying binoculars and a camera case. His name was Doctor David Hull, and he was an entomologist from a nearby university. I told him I thought the birdwing was the most beautiful butterfly in the world. Just then the creature flew up and disappeared in the treetops. "Too bad I'll never have one for my collection." (It's against the law to collect them without special permission.)

"You have a collection?"

"Max has one of the best amateur collections in the state," Dad told him.

Walkingstick

On our way back, Dr. Hull and I talked about my insect collection. We stopped to rest, and I sat down on a log.

"Max," Dr. Hull said, "take a close look at that long twig in front of you." I looked where he was pointing and saw the slender twig move as though swaying in a breeze. Except there was no breeze.

I looked closer and realized I was looking at the biggest **walkingstick** *(Order Phasmida)* I'd ever seen. It was nearly ten inches long. "He's three times longer than Speedy," I said.

"Speedy?" Dr. Hull asked.

"Max's pet walkingstick back home," Mom said. "He calls it Speedy as a joke because it almost never moves."

This walkingstick was so perfectly camouflaged that if Dr. Hull hadn't pointed it out I would never have spotted it. Its thin body and long antennae made it look exactly like just another twig on the tree branch.

I reached out and gently lifted it off the branch and set it onto Mom's arm. She smiled. She's used to holding Speedy. The insect never moved.

Walkingsticks eat plants, and Dr. Hull said in some parts of Australia they damage crops. But they're interesting pets and easy to take care of.

When we got back to town, we exchanged addresses. Dr. Hull said he had something he wanted to send me. I couldn't imagine what it was, but I thought it would be fun to keep in touch with an insect expert halfway around the world.

Atlas

Java is northwest of Australia. We flew to the capital city, Jakarta, and from there took a boat through the narrow strait between Java and Sumatra. We passed dozens of islands and a few active volcanos. There's a national park on the western tip of Java. We anchored near the mouth of the Cigenter river and took a small rubber boat upstream.

The sharp cries of colorful birds floated down to us from the treetops. We heard a tremendous thumping and crashing of branches. Dad put his hand on my shoulder and pointed, "It's a Javan rhinoceros, Max." The huge creature had come to the river to drink. He must have weighed three thousand pounds.

The strange surroundings made us want to talk in whispers. "This area gets more than one hundred inches of rainfall a year," Dad said. Thick vines hung from trees; flowers and plants grew everywhere. A hairy spider, about eight inches across, swam by our boat. We were near the shore when I noticed streams of water shooting up into the reeds and bushes.

"Look, Mom. There are archer fish underwater."

"What are they doing?"

"Hunting. They shoot down insects by spitting water at them."

"Look!" Dad whispered urgently. "On that tree trunk." He let the boat coast to a stop on the muddy river bank.

Not ten feet from us a giant **atlas moth** *(Order Lepidoptera)* was slowly spreading its wings. It was almost as big as the birdwing and its colors were even more incredible.

"How strange," Mom whispered. "The pattern on the tips of its wings looks like a snake's head."

"That's for protection," I said. "A bird would think twice before attacking something that looked like a big snake."

"See if you can get a picture of it," Dad said.

I did.

The Leaf That Wandered

"Look at that strange leaf," Dad said. We were sitting in the boat under the shade of a large tree that overhung the water.

"I don't think it's a leaf, dear," Mom answered. "Unless leaves can crawl."

On the tree trunk was one of the strangest insects I'd ever seen. I had to check my books to find out it was a **wandering leaf** *(Order Phasmida)*, a close relative of the stick insects, like the walkingstick I'd seen in Australia.

It was three or four inches long, and across its back and on each of its legs were growths that looked exactly like leaves—they even had stems and ribs! The disguise was perfect. Another example of how nature helps protect its creatures.

The Flower That Wasn't

Singapore, on the Malay Peninsula, was our next stop.

On a photo trip into the jungle, Mom stopped to look at some pink orchids. "Look, honey. These are just like the ones you gave me for our senior prom. They're beautif—eek!" She squealed and jumped back.

Dad laughed. " 'Beautifeek'? What's that?"

"Whatever it is, it's not a flower," she said.

"I think it's a praying mantis," I said, looking closer.

"We have praying mantises back home in the garden," Mom said, "but they don't look like that." This was a **Malaysian flower mantis** *(Order Mantidea)* and, including its front legs, was about six inches long. It was also colored like no mantis I'd ever seen before. It was mostly pink, and its wings, folded along its back, were a pale ivory color. But what was really weird were the wide flanges on each hind leg. They looked almost exactly like the petals of the pink orchid.

"*Mantis* is a Greek word meaning prophet," Dad explained. "It describes the way the insect stands with its forelegs raised up as if it's praying."

"And I'll bet I know what it's praying for," I said. "Something to eat." As we watched, a butterfly, attracted by the orchid, settled on one of the leaves. With a flash almost too fast to follow, the mantis reached out and grabbed it. Its strong spiked forelegs are hinged like jackknives, and the butterfly didn't stand a chance. Sudden, deadly violence is an everyday fact of life in the insect world.

We watched as the mantis settled down to neatly eat its victim alive. All mantises devour their food alive—and just about any insect they can get a hold of is food. Praying mantises are very helpful because they eat insects that hurt plants. Mom is always careful not to harm the ones she finds in our garden.

Army on the Move

We crossed the Indian Ocean and followed the sun toward Nairobi, Kenya, in East Africa. In Nairobi we joined a big-game hunt heading for Ngorongoro Crater in Tanzania. Naturally we were hunting with cameras.

Ngorongoro Crater is what's left of an ancient volcano that blew up. Over thousands of years it became a valley, and now it's one of the biggest game preserves in the world. We saw several prides of lions lazing under trees in the hot sun, and hundreds of wildebeests and zebras roaming the plains. Once we startled thousands of flamingoes wading and feeding in a nearby lake. As they took off they filled the sky like a huge, pink, noisy cloud.

Early the second morning I was startled awake by a scream from one of the nearby tents. I ran outside to see what was the matter.

Mom came running out a few seconds later. "Is a lion on the loose? Did a snake sneak into camp?"

"No. It's worse."

"Worse," Dad came out rubbing his eyes. "What could be worse?"

"Army ants. I think one of the guides got bitten. I hope it's not too bad."

Everyone cleared out of the way, and in the growing light of dawn we watched as a column of perhaps 150,000 of the giant **ants** *(Order Hymenoptera)* paraded by. The largest were almost an inch long and had awesome mandibles. Those powerful jaws can actually bite off pieces of flesh.

"I thought ants lived in hives like other social insects. Like in that ant farm I gave you," Dad said.

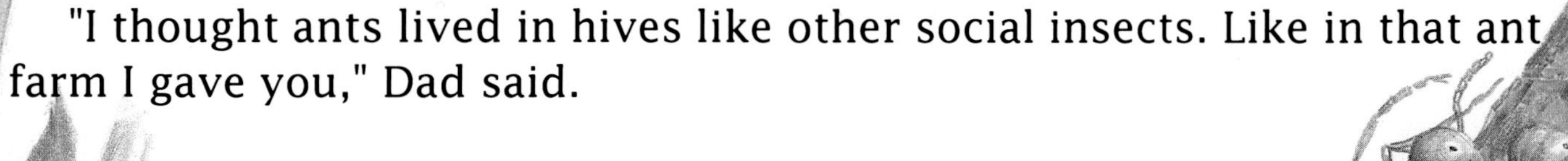

Like bees and wasps, most ants are social insects. That means they live together and help each other. "Not army ants, Dad. They eat so much they're always on the move looking for food. And they'll eat anything that gets in their way."

"What do you mean anything?" Mom asked nervously, looking at the guide who was being treated for several vicious bites.

"I've read that army ants have attacked goats and horses until there's nothing left but skeletons."

"Where are they all going?" Dad asked.

I explained that although it looks like they know where they're going, almost all army ants are blind and just follow the ant in front. They find food by smell, and if they run into an obstacle they just keep moving until they find their way around. "I read about a column that got separated from the main army and circled until the ants in front connected up with the ants in the rear. They just kept blindly circling around and around until they died of starvation."

I got out my collecting jar and scooped up several of the largest I could see, being very careful not to get bitten in the process. They'd make great additions to my collection.

The Tallest House

Later that day I got to see how another social insect lived.

Termites *(Order Isoptera)* aren't particularly big. At least most of them aren't. But the nests these mound termites build are huge.

"Dad! Take a picture of me next to this mound. Otherwise the kids back home will never believe me." I'm a little over five feet tall, and the mound towered over me by another fifteen feet.

Dad kicked at the base of the mound with his boot. "That thing is hard!"

The mounds are mostly made of sand mixed with juices from the termite's mouths. When they need to be removed for farming or building, big bulldozers can't tear them down and they usually have to be blown up with explosives.

"How many live in there?" Dad asked.

"Two million. Maybe more. Mostly soldiers and workers. There's only one king and queen. The queen's nothing but an egg factory. Her abdomen stretches out with eggs until she's as big as a breakfast sausage. Lots of termites eat wood and they can turn an abandoned shack into sawdust in no time."

"I don't think I want you taking any of those back home, Max," Mom said.

Battle in the King's Forest

"This used to be the private hunting ground of kings," Dad said. "William the Conqueror created the New Forest in 1079, and for hundreds of years it was against the law for anyone but royalty to hunt here. Now it's a public park."

We were near the southern coast of England. The ground was thick with leaves that crackled as we walked along.

"Max!" Dad said. "Look—a battle royal!"

And it was. But not between princes fighting for a kingdom. This was between two **European stag beetles** (*Order Coleoptera*) fighting for a female. And both were more than three inches long! Their dangerous-looking mandibles were waving around and hitting together as they tried to get hold of one another.

"Looks like you have to be careful handling one of those guys," Dad said. He was right. Their jaws are powerful enough to draw blood from a person's finger. The female, which is half the size of the male, was waiting off to one side.

Stag beetles spend almost all their lives as larva buried within tree trunks. After as long as five years, the adult beetle emerges to feed on sap from the tree bark and look for a mate.

"I don't think either one is going to win this fight," Mom said. But just then one managed to grab the other and lift it up in the air. Squirming and wiggling, it was carried over to the edge of the log and dropped off. It landed with a crunch on the dead leaves.

I was ready. I scooped it up in my jar before it could scuttle away.

Speed Demons and Bird Eaters

We were spending the last two weeks of our summer adventure in a small country called Belize on the Yucatan Peninsula south of Mexico. Belize has the longest barrier reef in the Western Hemisphere. The water was warm, clear and blue; the sun was bright; and the air was hot and humid.

While Mom and Dad were scuba diving, I got a chance to see one of the fastest insects in the world.

Just a few hundred yards from the beach was a large swamp. The tropical jungle grew up right to the water's edge. I found a comfortable rock and sat down quietly to wait.

As I got used to the stillness I realized it wasn't still at all. There were soft *plops* and ripples as fish broke the surface to eat small insects. Water striders glided by, hardly stirring the surface. I could hear the buzzing of gnats and mosquitoes (a good thing I had put on repellent!), and parrots called from somewhere deep in the jungle.

Suddenly, right in front of me, I saw movement almost too fast to follow. One second it was there, the next it was gone. There it was again! Gone again! Then I saw it. A **damselfly** *(Order Odonata)* finally came to rest on a reed close by. I stayed very still, not wanting to frighten it. It was beautiful. A very slender body, close to five inches long, and delicate, thin-veined,

blue-tipped wings that must have been six or seven inches wide.

It's easy to tell damselflies from dragonflies. Dragonflies' bodies are thicker and heavier and they always hold their wings straight out when resting, while damselflies hold theirs folded together above their abdomens.

Damselflies and dragonflies are some of the oldest types of insects on earth. There is a fossil of a dragonfly almost 300 million years old with a wingspan of twenty-seven inches—bigger than some birds! That makes it the biggest insect that ever lived.

As quickly as damselflies can move, dragonflies are much faster. Some have been seen flying at speeds of almost sixty miles an hour. They could pass us on the freeway. If you think that makes them tough to catch, you're right. I have only one in my whole collection.

Another reason they're hard to catch is that they can see better than any other insect. Each of their huge compound eyes has as many as thirty thousand separate facets.

They can see tiny bugs from as far away as forty feet. They always hunt while flying, and because they hang around swamps and ponds eating mosquitoes, gnats and flies, they're very helpful insects.

They live their whole lives near water. The eggs are laid underwater, and when they hatch, the larva, called nymphs or naiads, spend all their time underwater breathing through gills.

"Max!" Mom called from over by the beach. "Max! Where are you?"

"Over here, Mom."

She came around a twist in the trail. Just as she got to the rock I was sitting on, we both got the surprise of our lives. The biggest **spider** *(Arachnid)* I had ever seen ran out from under the rock and between Mom's feet.

She screamed. I leaped up and away. (I was a little embarrassed actually. I shouldn't have been frightened. Even though spiders aren't insects, I've studied them a little and I've handled tarantulas before. Their hair is soft and they never bite as long as you're careful. And even then their bite isn't much worse than a bee sting. But there's something about seeing a spider the size of a baseball mitt that kind of gives you the creeps.)

Dad came running up behind us. "Everything all right here—good grief! Is that a tarantula?" The spider was now several yards away, sitting back on its hind legs with its large fangs spread menacingly.

"I think it's a Bird Spider. And it's probably just as scared as we are." I said that so Mom would calm down, but it didn't seem to help much.

"Why do they call it that?" Dad asked.

"Because it eats birds, along with snakes and small mammals. A bird spider doesn't have many natural enemies because its hair is irritating to touch. And it doesn't spin webs as other spiders do to catch insects to eat. Instead it jumps on its prey and then injects a chemical that liquefies the victim's body so that it can suck out the contents."

When the Bird Spider realized we weren't a threat, it crawled slowly away into the thick underbrush. I wiped the sweat off my forehead. Probably just the heat, I thought.

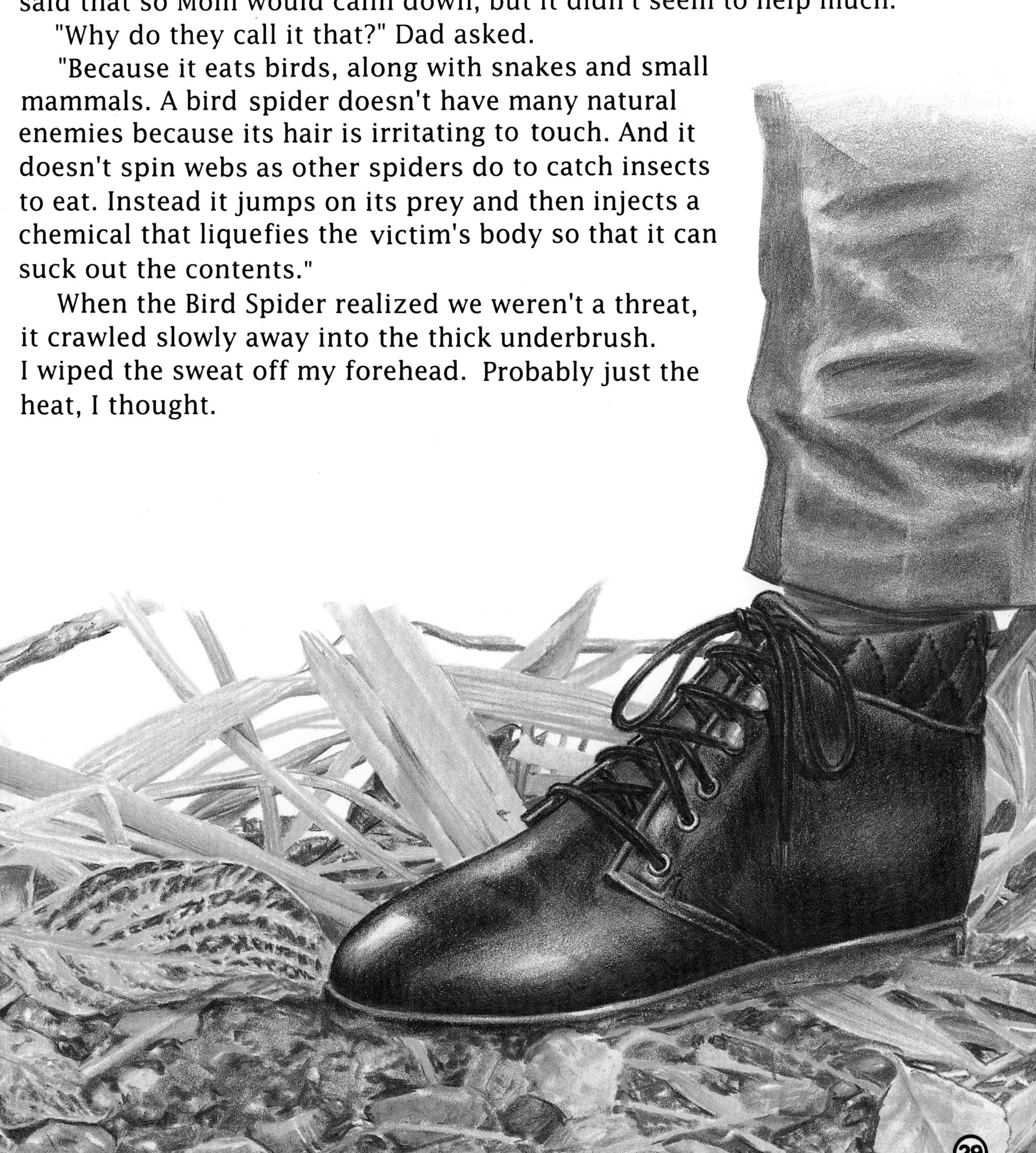

Surprise Package

What a summer! I had dozens and dozens of photographs and some great specimens to add to my collection. I'd had a great time with Mom and Dad. I'll bet we covered thirty thousand miles.

As much fun as we'd had, it was nice to be back home.

After we unpacked, Mom was going through the mail. "Max, there's a big package for you here."

"Where from?"

"Australia."

The package was wide and flat and heavily wrapped. Under the first layer of paper was a short note from Doctor Hull.

> Max, I enjoyed meeting you and hearing about your collection.Please write and let me know if the little present inside will add anything to it.
>
> Your friend,
> David Hull

Dad was smiling. "Well," Mom said, "open it."

My hands were shaking. I thought I knew what it was.

Sure enough, under the last layer of paper, its black velvet and bright green wings glinting in the light, was a beautifully mounted birdwing butterfly. I grinned. What a collection I was going to have now!

Glossary

Abdomen—The hindmost of an insect's three main body divisions.

Antenna (pl. antennae)—Feelerlike extremities located above the mouthparts on insect's head.

Arachnida—A large class of Arthropoda including spiders, scorpions, mites and related forms.

Arthropoda—A phylum consisting of invertebrate animals with jointed limbs. Includes insects, spiders, crustaceans and related forms.

Camouflage—Concealment by using a disguise. Color patterns or shapes that help insects hide from predators.

Chrysalis—The casing in which the pupa of butterflies and moths transforms into an adult.

Compound eyes—The major insect eyes located on the head and made up of many individual lenses.

Crustacea—A large class of Arthropoda comprising the majority of marine and freshwater arthropods such as lobsters, shrimp, crabs and related forms.

Entomologist—A scientist who specializes in the study of insects and related arthropods.

Exoskeleton—The external covering of an animal.

Fossil—The remains of an animal or plant of past geological ages that has been preserved in the earth's crust.

Head—On an insect, the region that bears the eyes, antennae and mouthparts.

Insect—The largest class of Arthropoda, which is made up of animals that have bodies consisting of a head, thorax and abdomen; three pairs of jointed legs; and other identifying features.

Instar—A stage in the life of an insect between successive molts.

Larva—The immature stage between the egg and the pupa in insects. Grubs, caterpillars and maggots are larva.

Mandible—One of the pair of mouthparts.

Metamorphosis—A change in form during the development of an insect.

Molt—To shed an outer layer of skin in a process of growth with the cast-off parts being replaced by new growth.

Naiad—The larval stage of damselflies and dragonflies.

Ocellus (pl. ocelli)—The simple eye of an insect.

Phylum—A major category of the animal kingdom containing various classes and orders.

Predator—An animal that eats other animals.

Proboscis—The extended beaklike mouthparts of an insect.

Pupa—The state between the larval and adult in insects. Usually a nonfeeding, inactive stage.

Thorax—The body area between the head and abdomen that contains the wings and legs.

Orders of Insects

COMMON NAME	ORDER	WORLD SPECIES
Aphids, Leafhoppers, etc.	Homoptera	32,000
Bark lice and Book lice	Psocoptera	1,100
Beetles and Weevils	Coleoptera	290,000
Caddisflies	Trichoptera	4,500
Chewing lice	Mallophaga	2,675
Cockroaches	Blattodea	3,500
Diplurans	Diplura	660
Dobsonflies, Lacewings, etc.	Neuroptera	4,600
Dragonflies and Damselflies	Odonata	4,950
Earwigs	Dermaptera	1,100
Fleas	Siphonaptera	1,370
Flies	Diptera	85,000
Katydids, Grasshoppers, etc.	Orthoptera	12,500
Mayflies	Ephemeroptera	2,000
Moths, Butterflies and Skippers	Lepidoptera	180,000
Praying Mantises	Mantidae	2,000
Proturans	Protura	375
Rock crawlers	Grylloblattodea	20
Scorpionflies and Allies	Mecoptera	350
Silverfish and Bristletails	Thysanura	580
Springtails	Collembola	6,000
Stoneflies	Plecoptera	1,550
Sucking lice	Anoplura	250
Termites	Isoptera	1,900
Thrips	Thysanoptera	4,000
True bugs	Hemiptera	23,000
Walkingsticks	Phasmida	2,000
Wasps, Ants and Bees	Hymenoptera	103,000
Webspinners	Embiidina	149
Zorapterans	Zoraptera	22

Bibliography

A Field Guide to the Insects of America North of Mexico
by Donald J. Borror and Richard E. White
Houghton Mifflin Company, Boston, 1970.

Simon & Schuster's Guide to Insects
by Dr. Ross H. Arnett, Jr. and Dr. Richard L. Jacques, Jr.
Simon & Schuster, Inc., New York, 1981.

Encyclopedia of Insects & Arachnids
by Maurice and Robert Burton
Finsbury Books, BPC Publishers, London, 1984.

The Audubon Society Book of Insects
by Les Line and Lorus and Margery Milne
Harry N. Abrams, Inc., New York, 1983.